ORION,
THE HUNTER

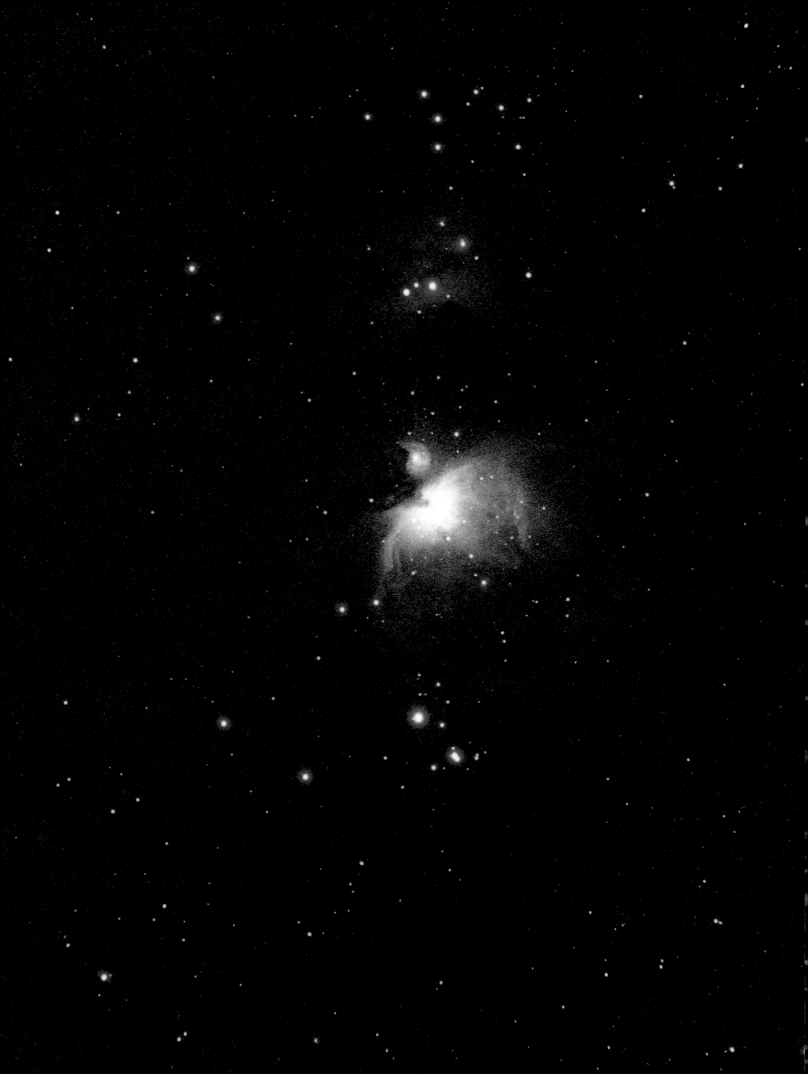

ORION,
THE HUNTER

NECIA H. APFEL

Clarion Books

NEW YORK

Also by Necia H. Apfel: *Voyager to the Planets*

Frontispiece: The Orion Nebula

Clarion Books
a Houghton Mifflin Company imprint
215 Park Avenue South, New York, NY 10003
Text copyright © 1995 by Necia H. Apfel

Text is 16/22 pt. Galliard Bold
Book design by Carol Goldenberg

For information about permission to reproduce selections from this book, write to
Permissions, Houghton Mifflin Company, 215 Park Avenue South,
New York, NY 10003.
Printed in Singapore

Library of Congress Cataloging-in-Publication Data
Apfel, Necia H.
Orion, the Hunter / by Necia H. Apfel.
p. cm.
Includes bibliographical references and index.
ISBN 0-395-68962-7
1. Orion (Constellation)—Juvenile literature. 2. Stars—Formation—
Juvenile literature. [1. Orion (Constellation) 2. Stars—Formation.] I. Title.
QB802.A64 1995
523.8—dc20   94-44268
CIP   AC

TWP 10 9 8 7 6 5 4 3 2 1

To my grandchildren—Benjamin, Hannah, Allison
And to their parents—Mimi and Andy, Steve and Barbara

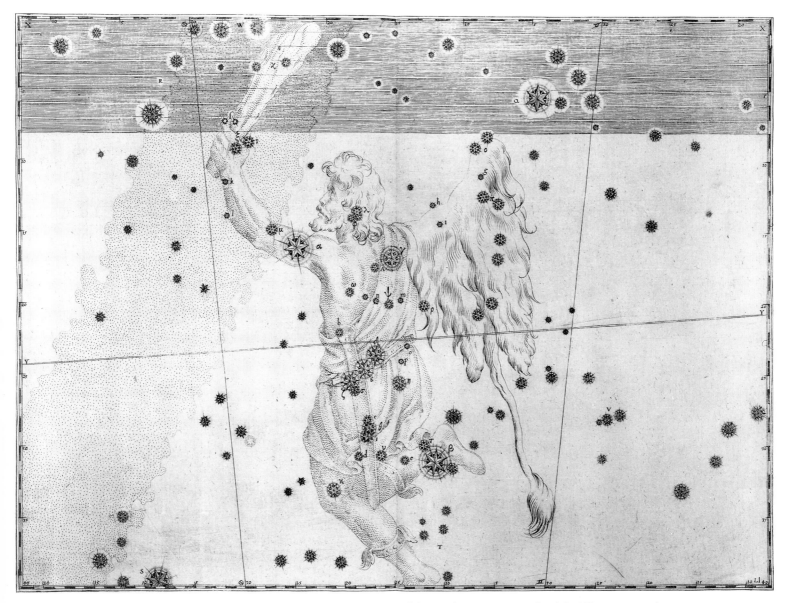

The constellation of Orion as depicted by Johann Bayer in his *Uranometria*, published in 1603.

Did you know there's a horse's head in the sky?
Would you like to see a place where stars are born?
Would you like to see a star that's as large as the
orbit of Mars?

All of these can be found in one region of the sky, in the famous constellation Orion. Today we use this ancient Greek name but all over the world, from very ancient times, this group of stars has had many different names. Many different stories have been told about the mythological character it is supposed to represent. According to Greek mythology, Orion was a giant hunter or warrior who accomplished many daring feats. In the sky, he is pictured facing the ferocious charge of Taurus, the bull, while his faithful dogs, Canis Major and Canis Minor, follow closely.

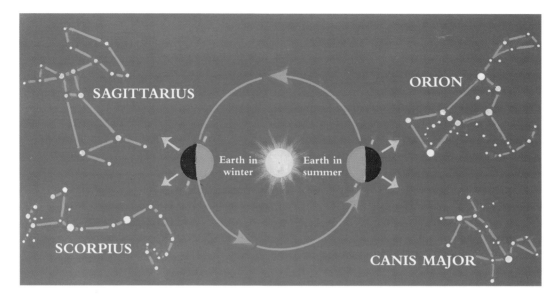

Throughout the year, different constellations are visible at night as the earth revolves around the sun. Orion, for example, is a winter constellation.

Orion is called a winter constellation because it is visible only from late October until April. The rest of the year it is in the sky only during daylight hours when, of course, our sun is too bright to allow any other stars to be seen. From December to March the constellation is easily observed throughout each clear night as it majestically crosses the southern sky. It is visible from every inhabited part of the earth during these months.

Where is Orion? It is one of the easiest constellations to find because it has so many bright stars, more than any other in the sky. On a dark, clear winter night, first look for three bright stars all in a row. They are close together and are equally spaced, one after another. This short line of stars represents Orion's belt.

The belt of Orion consists of a row of three bright stars. They are easily visible in the night sky, but most of the other stars in this photograph can be seen only through a telescope.

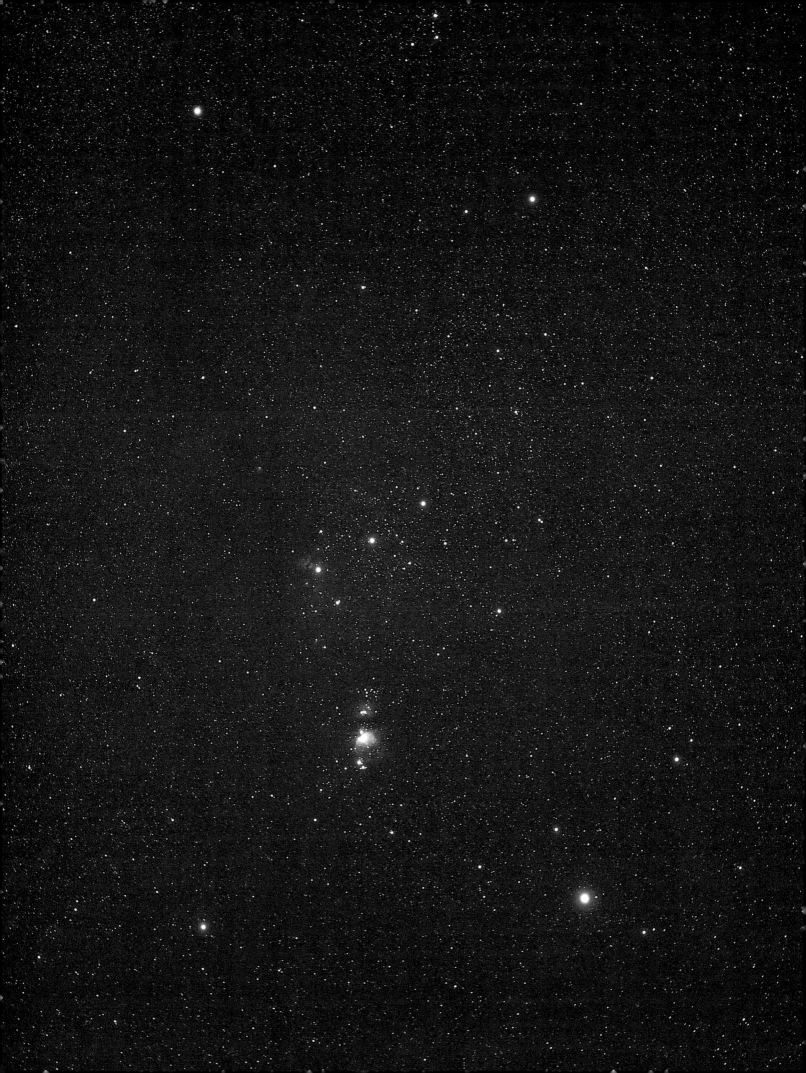

Once you have found Orion's belt, look a little higher in the sky for the two bright stars that represent his shoulders. Below the belt you will find two other bright stars that are his legs. These four stars form a large lopsided rectangle with the three-star belt at its center. If it is a very clear, moonless night, you may also be able to see some of the fainter stars in Orion. To the upper right of the large rectangle of bright stars is a curved line of six much dimmer ones. These outline the edge of Orion's large shield that he carries on one arm. The upraised club in his other hand is represented by five barely visible stars. And above the large rectangle is a group of faint stars as his head.

The constellation of Orion.

Below the easternmost star in Orion's belt, a ragged vertical line of fainter stars can be seen. This is the mighty hunter's sword. If you look carefully at the sword, you will note that about midway down there is a fuzzy bright patch of light. This can be seen much more clearly through a small telescope or a good pair of binoculars. It is not a star but instead is an enormous cloud of dust and gas particles, a nebula. It is called the Great Nebula of Orion or simply the Orion Nebula. It is over 1500 light years away. A light year is the distance that light travels in one year, about 6,000,000,000,000 miles. The light coming from the Orion Nebula has been traveling through space for more than 1500 years and is just reaching us now!

The Orion Nebula.

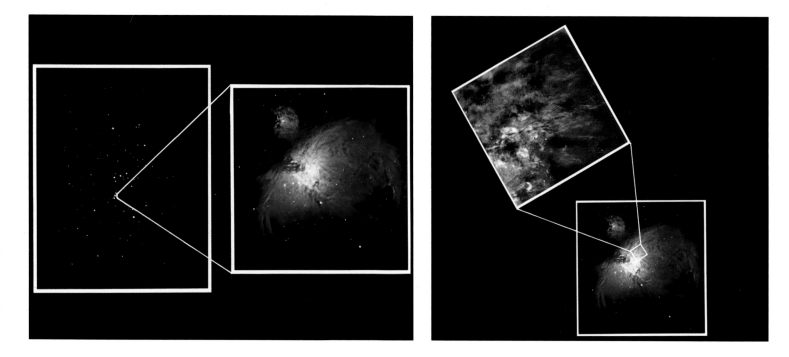

The Orion Nebula is an area that astronomers have been studying for many years. Within this enormous cloud are four very bright young stars that illuminate some of the dust and gas particles, causing them to shine in many different colors. The dust particles can be seen because they reflect the brilliant light of the stars in their neighborhood. The gas particles, which are much more numerous than the dust, can also reflect starlight. But if they are close enough to very hot stars like the four within the Orion Nebula, the strong ultraviolet light from these stars causes the gas atoms to fluoresce. Fluorescent bulbs give us light by the same process, although it is electricity, rather than the energy of ultraviolet starlight, that causes the gases in the bulbs to fluoresce.

A section of the Orion Nebula photographed by the Hubble Space Telescope's wide field camera. It shows the structure of a thin sheet of gas at the edge of the nebula. This sheet of gas has been compared to a rippled window curtain.

An artist's rendition of an infrared astronomical satellite (IRAS) in orbit around the earth. With the use of infrared telescopes, astronomers can penetrate the dust and gas and detect many very young stars that cannot be seen by optical telescopes.

You can see the four bright stars that make the Orion Nebula so magnificent by looking at the brightest section through a telescope or binoculars. However, unless the telescope is quite powerful, the stars won't appear very distinct. They form a little quadrangle that is called the Trapezium. This is the center of a dense cluster of more than three hundred very young, developing stars. Most are still too dim to be seen without powerful instruments, because they are still in the early stages of star formation. These are some of the approximately one thousand stars that have been created from the dust and gas in the nebula over the last million years. Some of these stars have already gone through their life cycles, and are long since gone, but many more are still in the process of being born. There is enough material for many thousands more stars to be formed. No wonder the Nebula is called a "stellar nursery."

Astronomers have found many young star clusters forming in other regions of the sky, but the Orion group is the closest to the earth. It is also one of the youngest, being at most 300,000 years old. It is estimated that the Trapezium stars themselves started to radiate only about 23,000 years ago. Over these many thousands of years their intense radiation has burned through the dust and gas in the nebula, enabling us to see their brilliance. Astronomically, these time periods are very recent. Our sun is about five to six billion years old.

The Trapezium consists of four very bright stars in the center section of the Orion Nebula.

The Pleiades star cluster in the constellation Taurus. Note the nebulosity still clinging to these relatively young stars.

By the time all the nebulous material is exhausted, many, many millions of years from now, the Orion Nebula will probably be gone and instead there will be a cluster of several hundred stars all moving together through space. A famous example of such a young cluster

that is no longer surrounded by much nebulous material is the Pleiades, which is fairly close to Orion in the sky. Although in the photograph only several dozen stars are visible, there are actually hundreds of stars in this beautiful cluster.

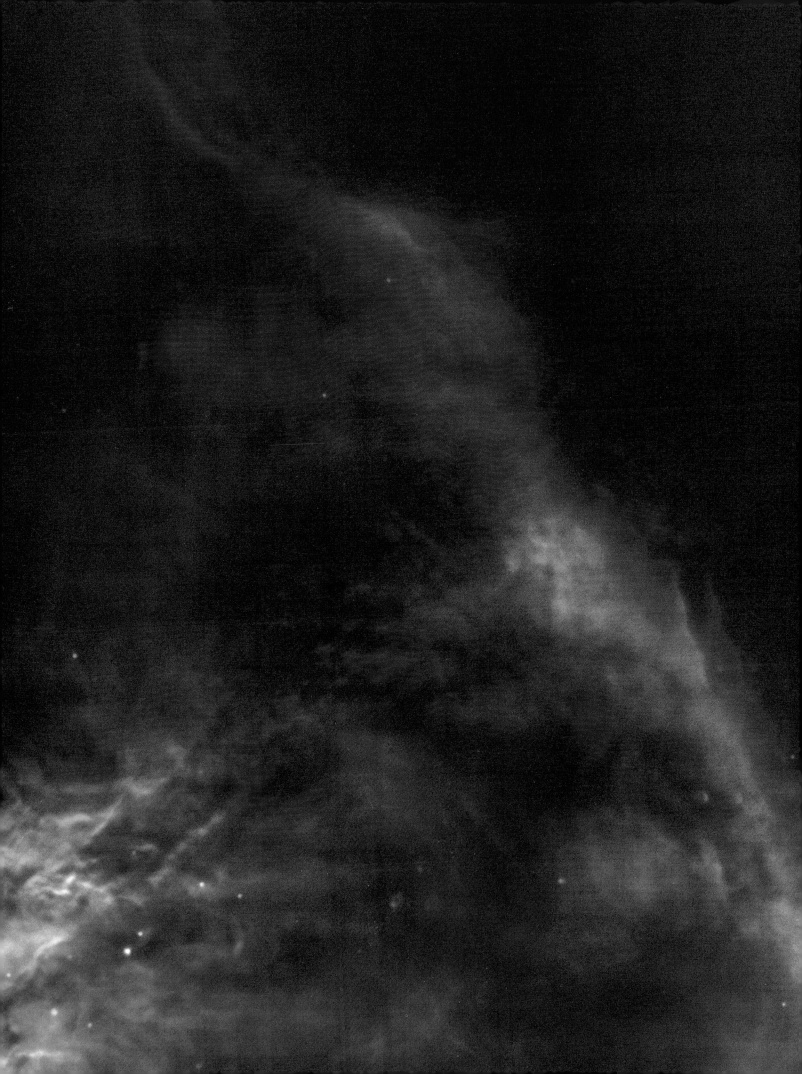

How do these stars form? The atoms and molecules of matter in the nebula are not spread uniformly over the region. Astronomers have discovered clumps or "globules" of material within the nebula that are much denser than their surroundings. How these globules started forming is still unknown because none of them have been detected at their very beginning. Shock waves from distant exploding stars or from the initial energy of a newly formed star may have pushed some of the particles closer together. Or perhaps different sections of the nebula collided, causing a disturbance that moved the particles. In a similar way, if you create a disturbance on the surface of a still pond of water, it will usually cause any leaves or debris to float closer together.

The Hubble Space Telescope discovered globules of matter within the dust and gas in the Orion Nebula. These will someday become stars.

Once a globule of atoms and molecules has formed within the nebula, the gravitational attraction between these particles increases because they are closer. This, in turn, enables them to pull closer still, becoming more dense than before. Their combined gravitation also allows them to pull other particles into the globule, making it grow more and more massive and dense. Its central temperature starts to rise as the pressure on its core increases. Eventually, the center region becomes extremely hot and nuclear fusion commences. To understand the enormous energy released by this process, imagine the force of thousands of hydrogen bombs detonating every second.

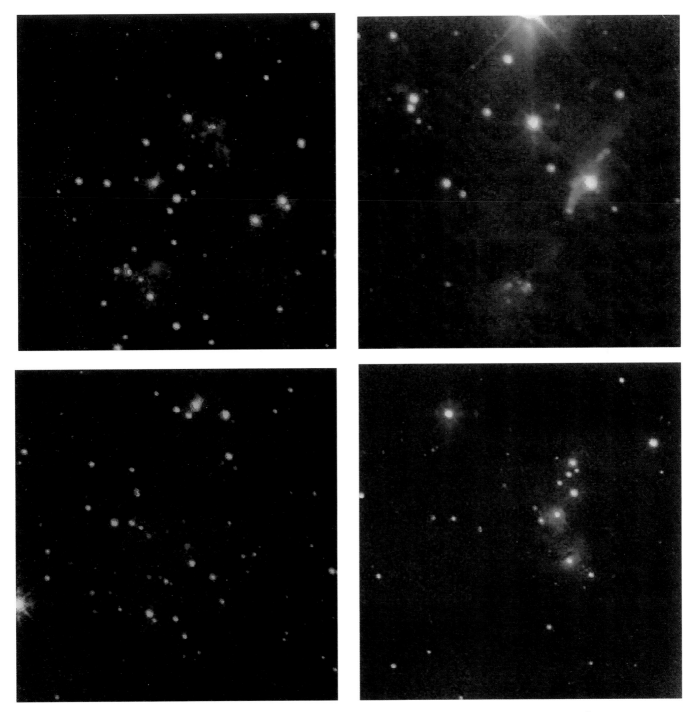

Using an infrared telescope, astronomer Karen Strom has found some of the youngest stars ever detected—about 500,000 years old. These newcomers were found deep within the dust and gas on the outskirts of the Orion Nebula. They are invisible to ordinary telescopes.

With nuclear fusion, the newborn star generates its own energy deep within its core as its abundant hydrogen gas is slowly converted into helium. When this energy works its way to the star's surface, we see it as starlight or, in the case of our sun, as sunlight. This process of star birth is continually happening in the Orion Nebula and in many other regions throughout the universe. All the stars you see at night started as small globules and are now generating their own energy deep inside their cores by nuclear fusion.

The Rosette Nebula in the constellation Monoceros. Central stars are pushing the gas and dust particles outward, creating the roselike configuration.

In 1992, astronomers using the Hubble Space Telescope made another discovery in the Orion Nebula. They found fifteen very young stars, each with a large disk of dust surrounding it. The dust is moving around each star too quickly to be pulled into it and instead it has spread out like a pancake with the star at its center. Astronomers believe that this dust may one day accumulate into larger bodies which will develop into planets and form new solar systems. It is the best evidence that many stars, other than our own sun, form planetary systems. It is believed that a similar disk of dust revolving around our sun formed into our solar system some four and a half billion years ago.

This Hubble Space Telescope photograph reveals a young star surrounded by a glowing disk of gas and dust (see inset). Planets, like those in our solar system, are believed to form from such disks.

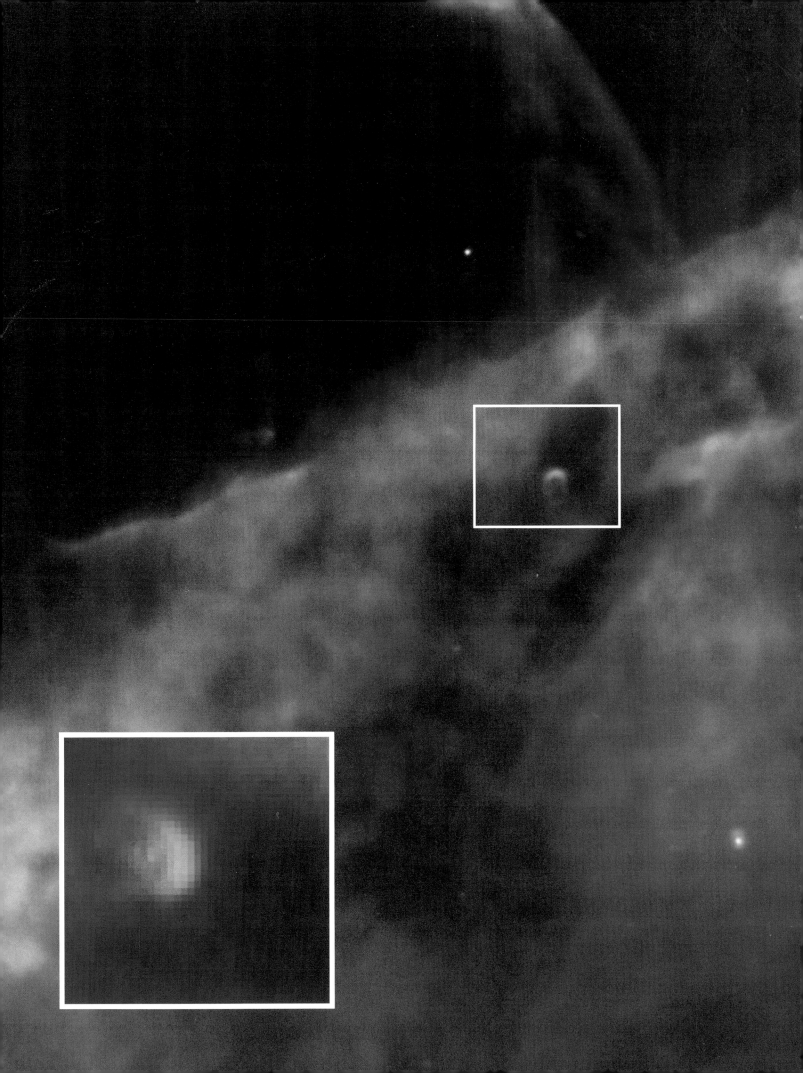

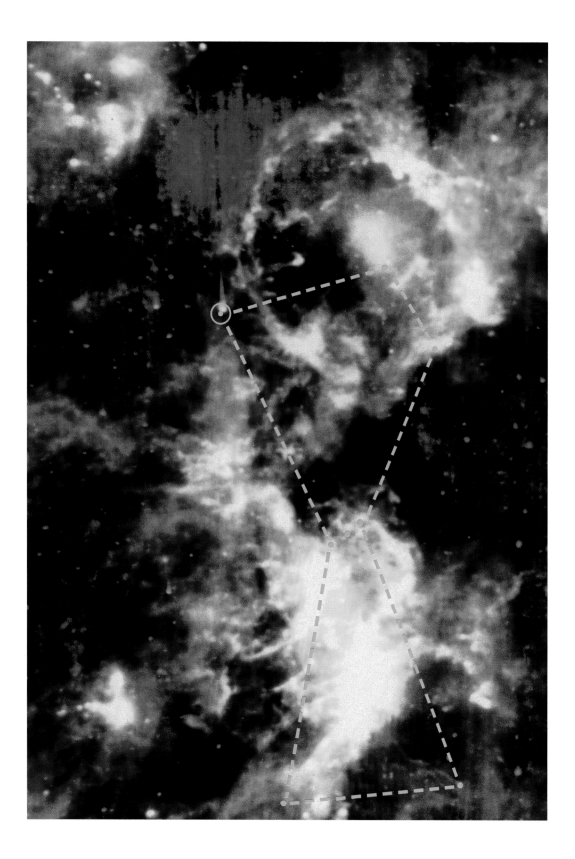

The Orion Nebula that we can see is about five or six light years across. But this is a very small fraction, perhaps about one-sixth, of the entire cloud of dust and gas. The rest of the nebula is spread out over much of the Orion constellation itself, covering about 20,000 times the diameter of our solar system. It consists of cool, dark dust and gas that does not emit any light. And although it is much less dense than any vacuum scientists can create here on earth, it is more than a thousand times denser than the thinly scattered dust and gas found in most of the space between the stars. Therefore, optical telescopes cannot see into its interior. It would be like trying to see objects in a very thick fog.

Although most of it is not illuminated, the Orion Nebula covers most of the Orion constellation. Only through the use of an infrared telescope can the full extent of the nebula be realized. The brightest areas are where stars are forming. The outline of the Orion constellation will help you locate the familiar stars.

However, there are other types of telescopes that can penetrate this dense cloud. Radio, infrared, x-ray, and microwave radiation passes through dense dust and gas with very little interference. This is because these types of radiation have wavelengths that are longer (radio and infrared) or shorter (x-ray and microwave) than light wavelengths. Different kinds of telescopes have been designed to detect and record each of these radiation wavelengths. Using them, astronomers have made many surprising discoveries in recent years.

Although most of the material between the stars is composed of single atoms, mainly hydrogen atoms, astronomers found that the nebula material is primarily made up of molecules. Molecules are combinations of atoms, and such material was not expected to be present in a nebula. However, more than six dozen types of molecules have been detected in the Orion Nebula, including water, carbon dioxide, carbon monoxide, and sulfur dioxide. More surprising was the discovery that the molecules making up what we now know to be the basic building blocks of life are also present in this seemingly inhospitable environment. It is no wonder that this enormous cloud is now called the Giant Molecular Cloud in Orion. It is truly a chemical factory, even producing some molecules that scientists can't make here on earth. In the photograph on page 30, the bright area at the bottom is the Giant Molecular Cloud.

A radio telescope. The Pleiades and the planet Venus shine in the night sky.

There is another famous nebula within the Orion constellation. It is called the Horsehead Nebula because of its very easily recognizable shape. It is too small to be seen with the naked eye or even through a small telescope; its distinctive configuration is revealed only through larger telescopes. And even then, long time exposures are necessary to show its unusual details. The Horsehead is located just below the lowest of the three stars in Orion's belt. This star is called Zeta Orionis. Its Arabic name is Alnitak. It is responsible for the strip of glowing gas lying far behind the Horsehead Nebula, causing it to be outlined against the bright background.

The Horsehead Nebula is faintly visible just below and a little to the left of the star Alnitak in Orion's belt.

The Horsehead is an example of a "dark nebula." No bright star is close enough to illuminate its dust and gas. We would not see it at all if it did not happen to lie in front of the bright nebula whose fluorescence is caused by Alnitak. There are perhaps about 400 light years between the Horsehead and Alnitak, with the closer Horsehead about 1200 light years from us. Its diameter is a little more than one light year across.

From the photograph you can see that the Horsehead is only a very small part of a much larger dark nebula that hides the light from the bright nebula behind it. Its dust particles, including those in the Horsehead itself, make up about 20 percent of its mass. They are about the size of those in cigarette smoke and are all in constant motion. But because of its enormous size and its great distance from us, the Horsehead shape never seems to change. However, many thousands of years from now it probably will not look the same.

The Horsehead Nebula.

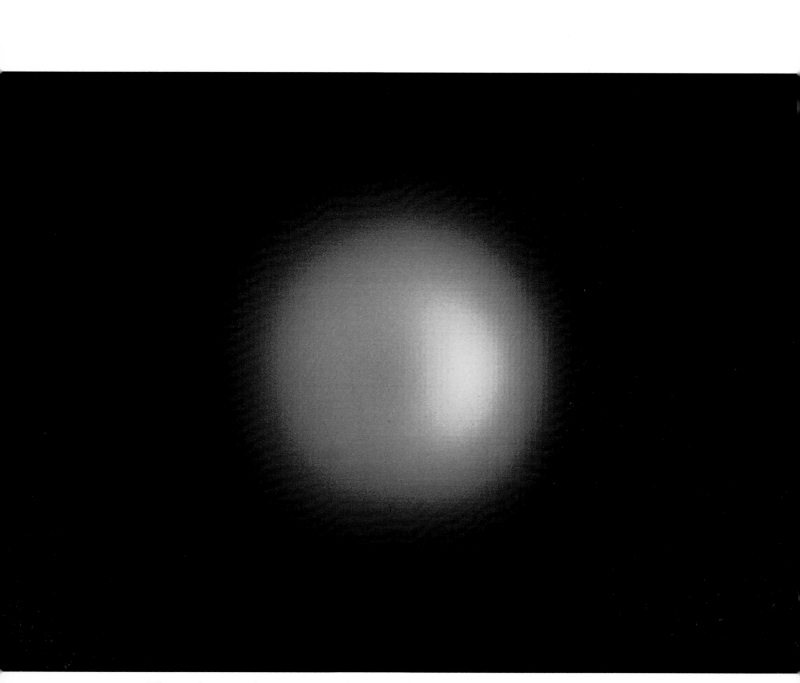

The red supergiant star Betelgeuse. The "hot spot" may be a huge column of rising gas. No other star except our sun has been seen so clearly.

Although the nebulae that are found in the Orion constellation attract a lot of attention, the region has many other features that astronomers are interested in. For example, look at the bright star in the upper left corner of the lopsided rectangle. Its name is Betelgeuse. Some people pronounce this name "beetle juice" and some say "beh-tel-jooz." Either way, it meant "the armpit of the giant" or "the arm of the central one" in the ancient Arabic language.

Betelgeuse is one of the largest stars known. Its size changes periodically so that its diameter ranges from 550 to 920 times that of our sun, which is about 850,000 miles across. If Betelgeuse were where the sun is now, it would extend out as far as the orbit of Mars and, at times, farther. Because it is so much larger than the sun, it is enormously brighter. But its brilliance varies also, ranging between 7600 and 14,000 times that of the sun. Can you imagine what it would be like if our sun's brightness varied by so great an amount over a period of only a few years? We can get an idea of how very bright this star really is when we realize that it is 520 light years away.

You will note that Betelgeuse, unlike most of the other stars in the sky, has a reddish tint to it. We call such stars red supergiants. These stars are very old and are in the last stages of their lives. They once were much smaller but expanded to enormous sizes when the energy-producing process in their central cores underwent radical changes. Their outer layers of gases, although still very hot, have thinned out and are no more than one ten-thousandth the density of the air we breathe every day here on earth. Such stars have sometimes been called "red-hot vacuums."

Compare Betelgeuse with the bright star on the lower left of the rectangle. This star is called Rigel, meaning "the left leg of the giant." It is a blue-white giant star, the seventh brightest in the sky. Unlike Betelgeuse, Rigel is a relatively young star and is only about fifty times wider than the sun. However, it is much more massive than Betelgeuse or the sun and is using up its energy-producing fuel at a very rapid rate. Its luminosity is fifty thousand times greater than that of the sun and its internal temperature is about six times hotter. Like a large bonfire that burns very quickly and brightly, Rigel will not last as long as most stars but will outshine them during its short life. In a few million years, Rigel will become a red supergiant like Betelgeuse.

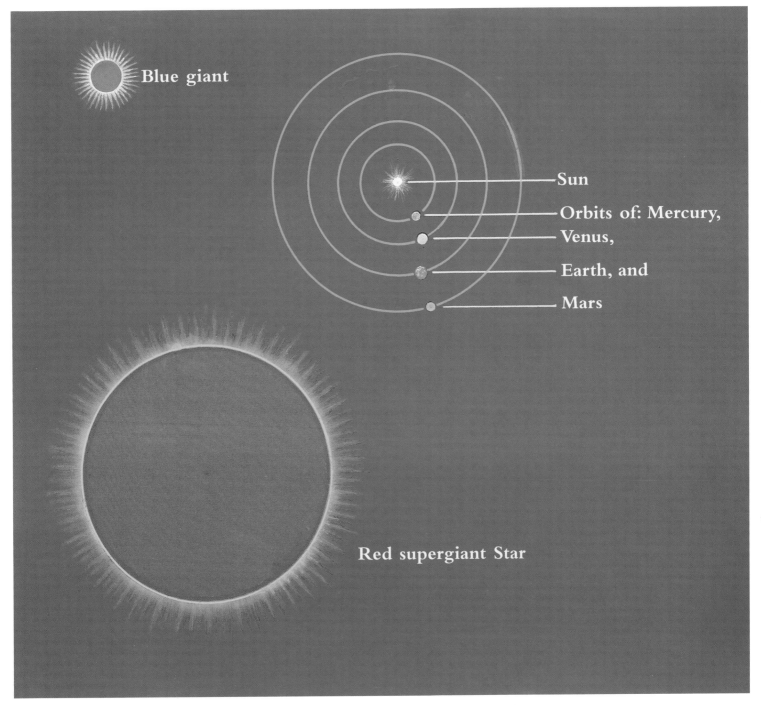

Blue giant

Sun

Orbits of: Mercury,

Venus,

Earth, and

Mars

Red supergiant Star

A comparison of some stellar sizes with the orbits of the four innermost planets in our solar system. Betelgeuse is a red supergiant; Rigel is a blue giant star.

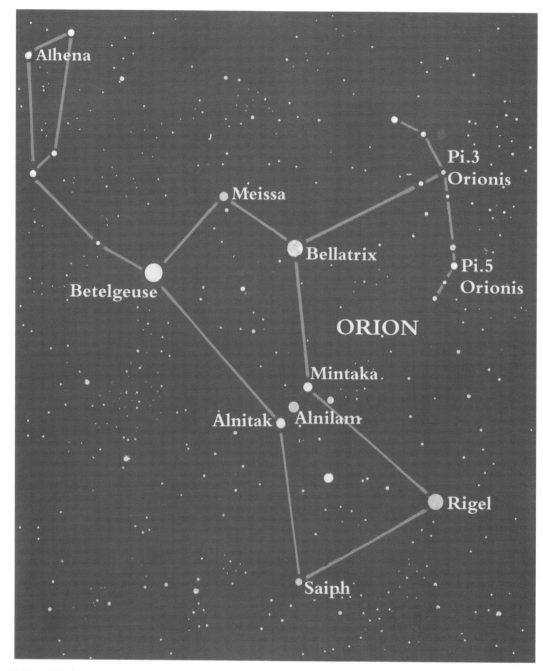

The brightest stars in the constellation of Orion. The sizes of the stars in this diagram represent their relative brightness as they appear in the sky.

Less than one star in a thousand is as hot and luminous as Rigel and none of the 300 nearest stars to us are as bright. Orion is unusual because it boasts not only Rigel but also five other blue-white supergiants, more than any other constellation. There is Bellatrix, Orion's left shoulder, and Saiph, his right knee, plus the three stars in his belt: Mintaka, Alnilam, and Alnitak. These bright stars as well as most of the hundreds of dimmer ones are all believed to have been born out of the same nebula and are moving together through space.

Halley's Comet. The debris from this comet creates the Orionid meteor shower every October.

A sky map upon which the trails of 42 meteors were traced during one hour of an Orionid meteor shower.

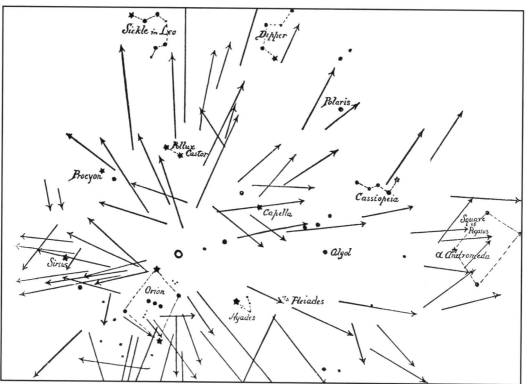

Although it has nothing to do with stars in the Orion constellation, every October 20 to 22 the Orionid meteor shower can be seen in this region of the sky. The meteors, which can occur from 10 to 70 times an hour, are brief streaks of light in the sky. They are caused by very rapidly moving bits of matter burning up from friction with air molecules as they enter the earth's upper atmosphere. These particles of matter are debris left behind by Halley's Comet, which was last seen during the 1985-86 winter. In late October the earth in its orbit around the sun enters this stream of comet debris, colliding with some of it as it moves along. The shower is named Orionid because the meteors appear to come from Orion.

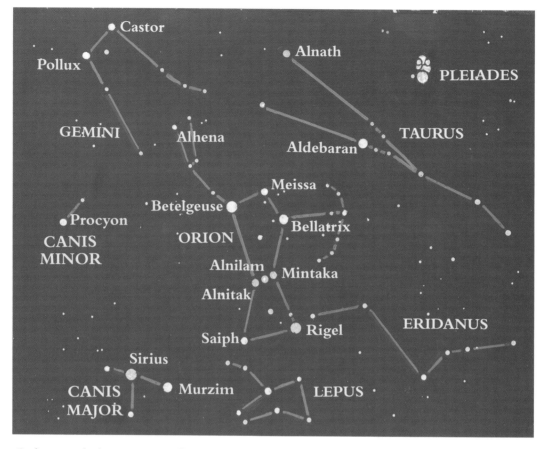

Orion and the surrounding bright stars and other constellations. The sizes of the stars in this diagram represent their relative brightness as they appear in the sky.

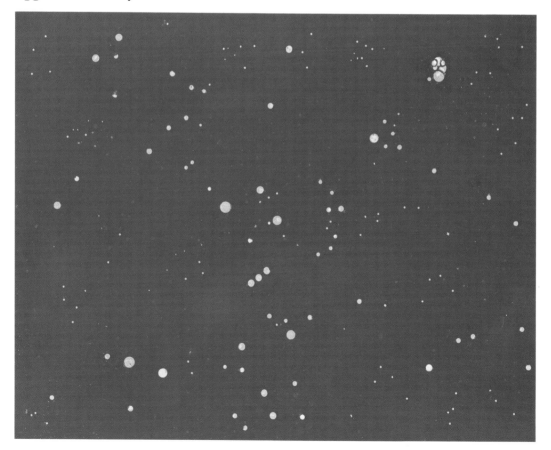

Surrounding Orion lie several other constellations with stars as bright or brighter than those seen in the mighty hunter. Sirius, the dog star in Canis Major, is the brightest in the northern sky and is one of the closest stars to earth. Canis Minor has Procyon which lies to the left of Sirius. With Betelgeuse, these two stars form a nearly equilateral winter triangle. On Procyon's upper left are Castor and Pollux, the twins of Gemini. And at Orion's upper right is Aldebaran, the orange-red eye of Taurus, the bull. Nearby is the Pleiades cluster. A winter journey through the sky with a pair of binoculars or small telescope will reveal many of the beautiful wonders of this region. The time is well spent.

Picture Credits

The Adler Planetarium, Chicago, Illinois: 6

Sadao Nojima: frontispiece, 9, 10, 34

Space Telescope Science Institute and National Aeronautics and Space Administration: 14, 15, 23, 29

Infrared Processing and Analysis Center, JPL: 16, 30

Lick Observatory: 18-19

Karen Strom, University of Massachusetts and National Optical Astronomy Observatories: 25

Hale Observatories: 20-21, 26-27

C. R. O'Dell (Rice University): 29

Bill and Sally Fletcher: 37

David Buscher, Chris Haniff, John Baldwin, and Peter Warner, Mullard Radio Astronomy Observatory: 38

Diagrams on pages 41, 42, 44, and 46 by George Buctel.

Index

Aldebaran, 47
Alnilam, 43
Alnitak, 35–36, 43

Bellatrix, 43
Betelgeuse, 38–42, 47

Canis Major, 7, 47
Canis Minor, 7, 47
Castor, 47

Fluorescence, 14

Gemini, 47
Giant Molecular Cloud, 30, 32
Great Nebula of Orion. *See* Orion Nebula

Halley's Comet, 44–45
Horsehead Nebula, 34–37

Infrared astronomical satellites (IRAS), 16

Light years, 12

Meteors, 44–45
Mintaka, 43
Monoceros, 26–27

Nebulae
    bright and dark, 36
    formation of stars, planets, and solar systems in, 17–28
    Great Nebula of Orion. *See* Orion Nebula
    Horsehead Nebula, 34–37
    Rosette Nebula, 26–27
Nuclear fusion, 24, 27

Orion (constellation), 6, 8–12, 35, 39–40, 43, 46–47
Orion, myth of, 7
Orion Nebula, 12–15, 31–32

age of stars in, 18, 25
Giant Molecular Cloud in, 30, 32
molecule types in, 32
planet formation in, 28
size of, 31
star formation in, 17–20, 23–27
Orionid meteor showers, 44–45

Planets, formation of, 28
Pleiades, 20–21, 33, 47
Pollux, 47
Procyon, 47

Rigel, 40–43
Rosette Nebula, 26–27

Saiph, 43
Sirius, 47
Solar systems, formation of, 28

Stars
    blue-white giant, 40–41, 43
    formation of, 17–27
    red supergiant, 38–41
Sun, 18, 27–28, 38–40

Taurus, 7, 20, 47
Telescopes, 16, 25, 28, 31–33
Trapezium, 17–19

Venus, 33

Winter constellations, 8

Zeta Orionis. *See* Alnitak